Animals Live Here

Life in the Ocean

By Connor Stratton

level 2 little blue readers

www.littlebluehousebooks.com

Little Blue House is distributed by North Star Editions:
sales@northstareditions.com | 888-417-0195

Produced for Little Blue House by Red Line Editorial.

Photographs ©: IreneCorti/iStockphoto, cover; georgeclerk/iStockphoto, 4, 24 (top left); VitalyEdush/iStockphoto, 7 (top); RamonCarretero/iStockphoto, 7 (bottom); pkphotoscom/iStockphoto, 8–9, 24 (bottom left); adamkaz/iStockphoto, 11; Hoatzinexp/iStockphoto, 12; Fgorgun/iStockphoto, 15 (top); Yurkovski/Shutterstock Images, 15 (bottom); Vlad61/Shutterstock Images, 17 (top); Arunee Rodloy/Shutterstock Images, 17 (bottom); UWPhotog/iStockphoto, 18–19, 24 (bottom right); kerkla/iStockphoto, 20–21; Maui Topical Images/Shutterstock Images, 23, 24 (top right)

Library of Congress Control Number: 2019908641

ISBN
978-1-64619-023-2 (hardcover)
978-1-64619-062-1 (paperback)
978-1-64619-101-7 (ebook pdf)
978-1-64619-140-6 (hosted ebook)

Printed in the United States of America
Mankato, MN
012020

About the Author

Connor Stratton enjoys visiting the ocean, spotting new animals, and writing books for children. He lives in Minnesota.

Table of Contents

dolphin

Ocean Animals

Many different animals live in the ocean.

Dolphins live in the ocean.

Rays live in the ocean.

They have large fins.

Sharks live in the ocean.

They have sharp teeth.

ray

shark

Seals live in the ocean.

They have long whiskers.

They can hold their breath for a long time.

seal
whisker

Jellyfish live in the ocean.

They have tentacles.

They have no brain.

They have no eyes.

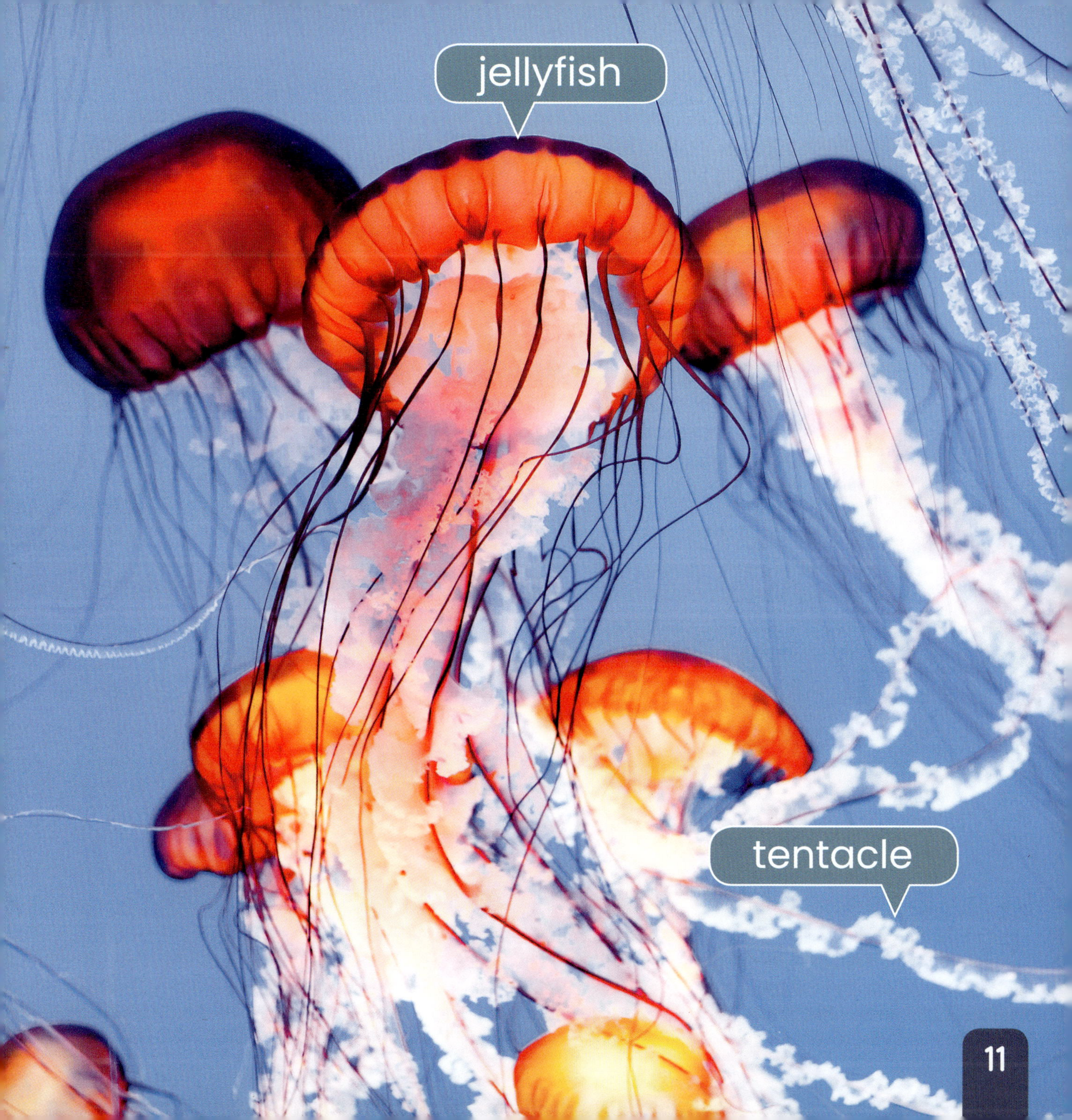
jellyfish
tentacle

fish

Many Fish

Fish live in the ocean. Some fish swim in large groups.

Some fish are yellow, and some fish are blue. Some fish are orange and white.

Some fish have stripes.

Some fish have spots.

stripe

spot

whale

Different Sizes

Whales live in the ocean.
Whales are the biggest
animals in the ocean.

Shrimp live in the ocean.

Shrimp are small.

They have shells.

shrimp
shell

Turtles live in the ocean.
They have shells.
Turtles can be big
or small.

turtle

Glossary

dolphin

turtle

seal

whale

Index